ANIMALS OF THE TUNDRA

POWERFUL POLAR BEARS

THERESA EMMINIZER

PowerKiDS press.
New York

Published in 2022 by The Rosen Publishing Group, Inc.
29 East 21st Street, New York, NY 10010

First Edition

Portions of this work were originally authored by Maeve T. Sisk and published as *Polar Bears*. All new material in this edition authored by Theresa Emminizer.

Editor: Jill Keppeler
Book Design: Michael Flynn

Photo Credits: Cover, pp. 1, 7, 15 (walrus) Paul Souders/Stone/Getty Images; (series background) Stefan Sorean/Shutterstock.com; p. 5 Robert Postma/Design Pics/Getty Images; p. 9 Chase Deckker Wild-Life Images/Moment/Getty Images; p. 11 Chasing Light - Photography by James Stone james-stone.com/Moment/Getty Images; p. 13 HuntedDuck/iStock/Getty Images; p. 15 (polar bear) schuie/iStock/Getty Images; p. 17 SeppFriedhuber/E+/Getty Images; p. 19 Lynn_Bystrom/iStock/Getty Images; p. 21 Ilya Akinshin/Shutterstock.com.

Library of Congress Cataloging-in-Publication Data

Names: Emminizer, Theresa, author.
Title: Powerful polar bears / Theresa Emminizer.
Description: New York : PowerKids Press, [2022] | Series: Animals of the tundra | Includes index.
Identifiers: LCCN 2020021705 | ISBN 9781725326439 (library binding) | ISBN 9781725326415 (paperback) | ISBN 9781725326422 (6 pack)
Subjects: LCSH: Polar bear–Juvenile literature. | Tundra animals–Juvenile literature.
Classification: LCC QL737.C27 E5334 2022 | DDC 599.786–dc23
LC record available at https://lccn.loc.gov/2020021705

Manufactured in the United States of America

CONTENTS

Home in the Tundra

Polar bears are some of the biggest meat-eating animals on land. They make their home in the cold, snowy **tundra**. They spend a lot of their time on sea ice, traveling to look for food.

A Big Body

Male polar bears are bigger than **females**. They can weigh up to 1,600 pounds (725.8 kg)! Their body is big, but their head is small. They also have a tiny tail that's only about 3 to 5 inches (7.6 to 12.7 cm) long.

Not White!

Polar bears aren't really white. Their fur is clear. It looks white because it **reflects** light. Their skin under the fur is black. This helps the sun keep them warm even when it's very cold.

Fur Feet

The fur on a polar bear's feet keeps it warm even in the snow. This fur also helps the bear walk on ice. Its feet have uneven skin and long claws to help too.

Sense of Smell

Seals are a polar bear's favorite meal. If a seal is on the ice, a polar bear can smell it from up to 20 miles (32.2 km) away! Polar bears can even smell seals through layers of ice and snow.

Meat Lovers

Polar bears only eat meat. They hunt seals, sea lions, and walruses. Sometimes, they even kill and eat whales! They can swim as fast as 6 miles (9.7 km) an hour for hours at a time!

walrus

Polar bears spend more than half their time hunting seals. A bear will only catch one or two of every 10 seals it hunts. As Arctic ice melts because of **climate change**, they have to travel even farther.

Baby Bears

Mother polar bears have cubs during the winter. They usually have two babies at a time. They keep the cubs safe and find them food. Young polar bears will stay with their mother for a number of years.

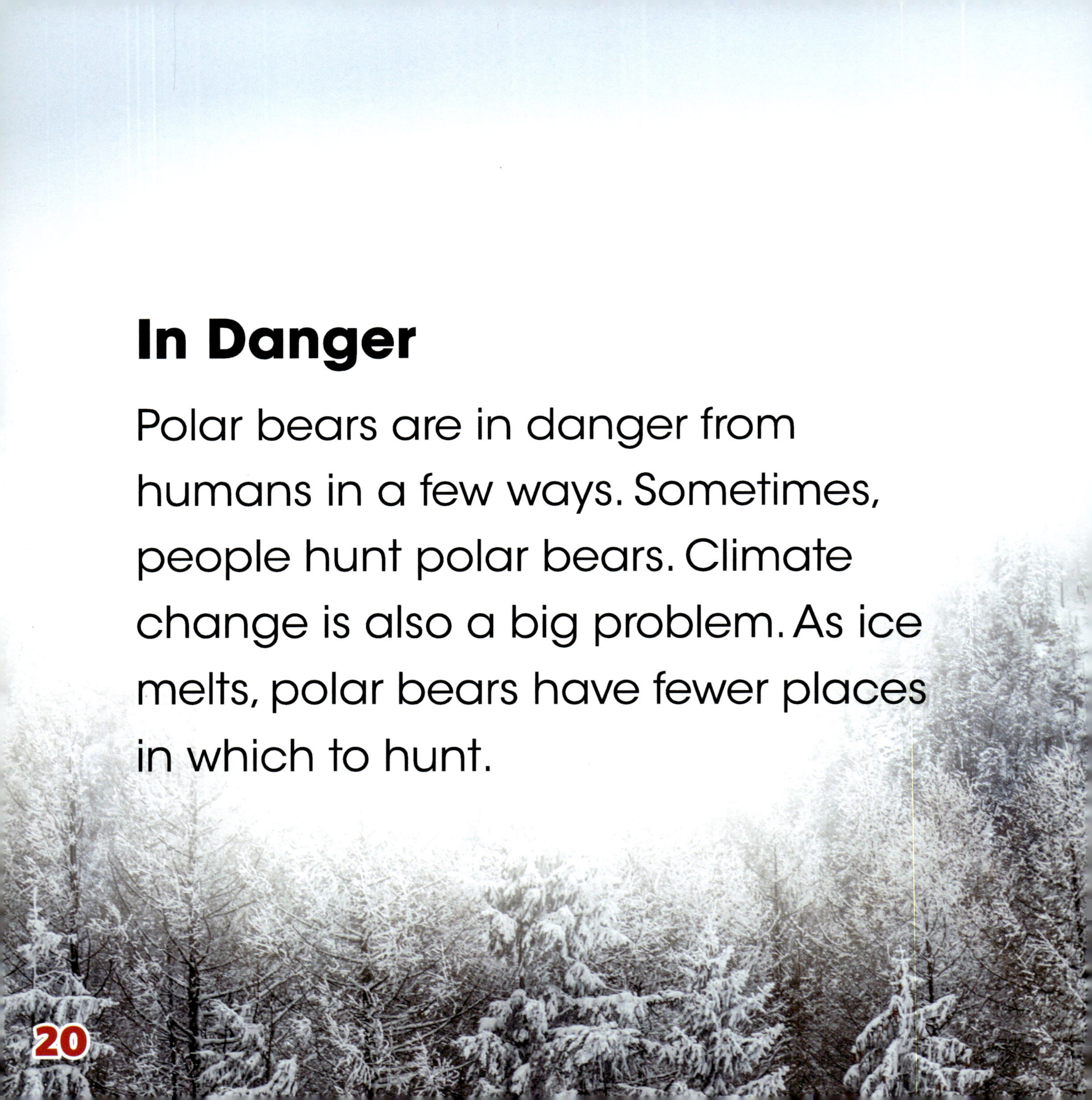

In Danger

Polar bears are in danger from humans in a few ways. Sometimes, people hunt polar bears. Climate change is also a big problem. As ice melts, polar bears have fewer places in which to hunt.

Fancy Features

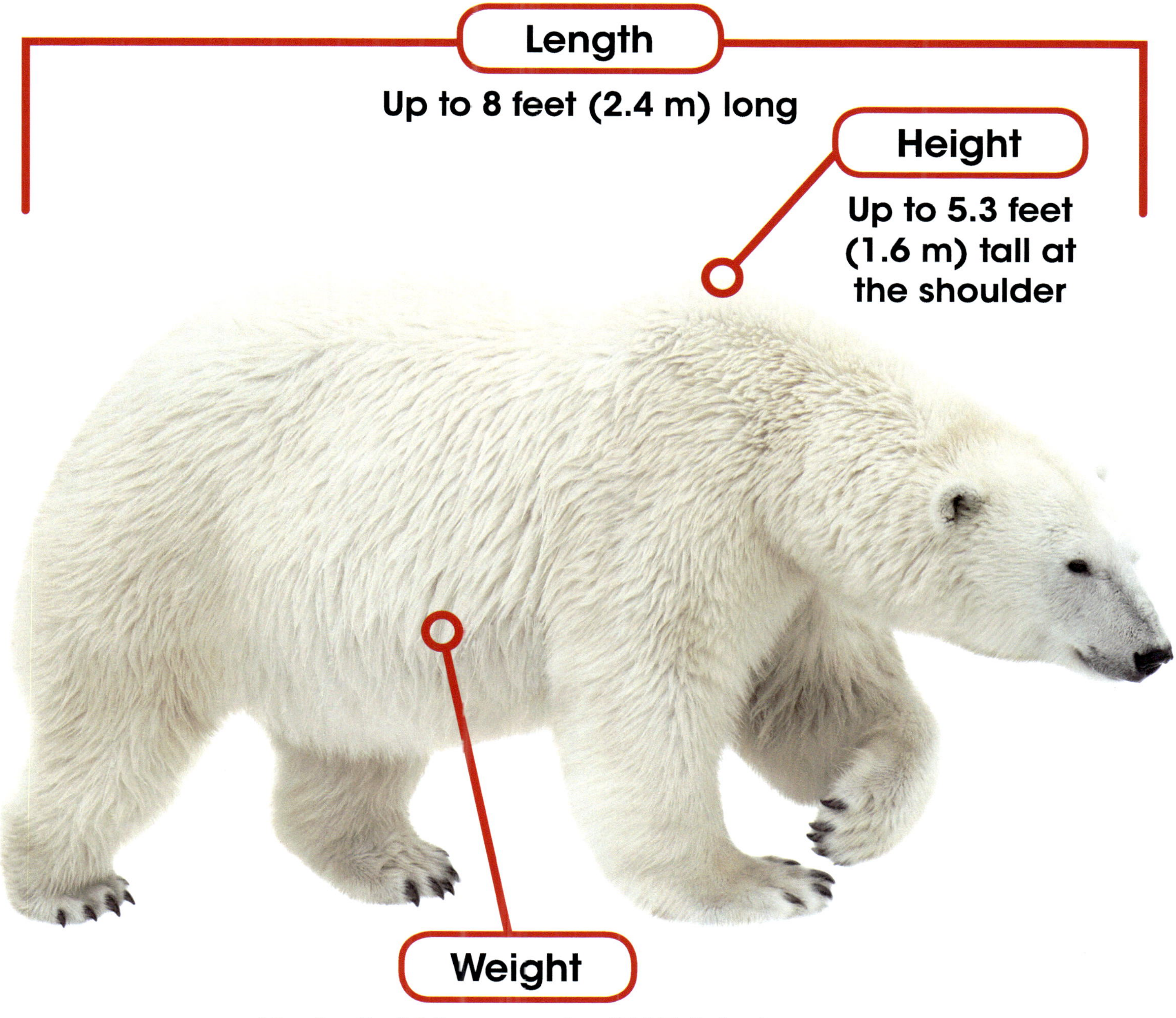

GLOSSARY

climate change: Change in Earth's weather caused by human activity.

female: A girl.

male: A boy.

reflect: To throw back light, heat, or sound without absorbing it, or taking it in.

tundra: Cold northern lands that lack forests and have permanently frozen soil below the surface.

FOR MORE INFORMATION

WEBSITES

Polar Bear
www.nationalgeographic.com/animals/mammals/p/polar-bear/
Watch videos about polar bears and learn more intriguing facts about these endangered animals on this National Geographic website.

Polar Bears
animals.sandiegozoo.org/animals/polar-bear
The San Diego Zoo presents this information-filled website about polar bears and their lives.

BOOKS

Marsh, Laura. *National Geographic Readers: Polar Bears*. London, UK: Collins, 2017.

Roome, Hugh. *Polar Bears.* New York, NY: Children's Press, Scholastic Inc., 2018.

Publisher's note to parents and teachers: Our editors have reviewed the websites listed here to make sure they're suitable for students. However, websites may change frequently. Please note that students should always be supervised when they access the internet.

INDEX